FLORA OF TROPICAL EAST AFRICA

OXALIDACEAE

CHRISTINE H. S. KABUYE
(East African Herbarium)

Annual or perennial herbs or subshrubs, rarely shrubs or trees. Leaves basically alternate, exstipulate or stipulate, simple or, more usually, digitately or pinnately compound. Flowers in axillary cymes or in pseudumbels or solitary, regular, hermaphrodite. Sepals 5, free, with imbricate aestivation. Petals 5, free or somewhat adherent near the base. Stamens 10(15), 2(3)-seriate; filaments fused below to form a tube with alternate short and long stamens; anthers 2-thecous, versatile, opening inwards. Ovary superior, 5-locular; styles 5, rarely 1; stigmas capitate, entire, bifid or laciniate; ovules 1–many in each locule; placentation axile. Fruit a capsule dehiscing loculicidally, or a berry. Seeds with the endosperm fleshy or absent; embryo straight.

A family of 7 genera and about 900 species.

Only two genera, *Oxalis* and *Biophytum*, occur naturally in Africa. *Averrhoa* L., which is supposedly a native of tropical America and the West Indies, is found in cultivation in isolated areas. The commonest species, *A. carambola* L., is a small densely branched tree, the leaves imparipinnate with 2–5 pairs of subopposite ovate to lanceolate leaflets, the flowers small in axillary panicled cymes, the fruit a fleshy yellow berry with 5 acutely angled lobes. The other species, *A. bilimbi* L., differs from *A. carambola* by having 5–17 pairs of oblong to linear-lanceolate leaflets and rounded lobes to the berry. The berries are very acid but can be eaten raw or cooked and sometimes made into preserves. *A. carambola* is cultivated at Amani and Mombo in Tanganyika, on the island of Zanzibar, in the Botanic Gardens, Entebbe, and the wetter parts of western Uganda. *A. bilimbi* is found in cultivation at Amani.

Fruit a capsule; herbs:
  Valves of the capsule persisting around the central column; leaves digitately or pinnately (1–)3-foliolate . . . . 1. **Oxalis**
  Valves of the capsule finally spreading; leaves paripinnate, usually sensitive to touch . . . 2. **Biophytum**
Fruit a berry; trees . . . . **Averrhoa** (see note above)

1. OXALIS

L., Sp. Pl.: 433 (1753) & Gen. Pl., ed. 5: 198 (1754)

Acetosella [Moehr., Hort. priv.: 4 (1736)] O. Ktze., Rev. Gen. 1: 90 (1891)

Caulescent or acaulescent perennial or annual herbs, bulbous or not. Leaves radical or cauline, alternate, digitately or pinnately (1–)3-foliolate, rather sensitive to light and drooping at night, petiolate, sometimes with membranous stipules adnate to the petiole. Flowers pedicellate, subtended by bracts, in pseudumbels or cymes or solitary, usually trimorphic heterostylous (largely homostylous in *O. corniculata*). Sepals sometimes with apical calli. Petals contorted in bud, falsely connate near the base. Stamens 10,

2-seriate, with 5 outer short filaments opposite the petals and 5 inner long filaments opposite the sepals; filaments usually fused at the bottom to form a tube and sometimes the long filaments each with an appendage at the back; anthers opening by slits. Ovary of 5 fused carpels, with 1–many ovules in each carpel; stigma terminal, capitate, 2-fid or laciniate. Fruit a capsule, with the valves persisting around the central column. Seeds brown, covered by a white fleshy integument, which bursts elastically by contraction expelling the seeds when ripe.

A cosmopolitan genus of some 800 species, mostly in America and South Africa.

1.	Leaves pinnately 3-foliolate; plant erect	1. *O. barrelieri*
	Leaves digitately 3-foliolate	2
2.	Bulbs absent; stems branching, creeping or ascending; corolla yellow	3
	Bulbs present; stems, if present, erect; corolla pink or purple to bluish	4
3.	Stipules present, adnate to base of petiole	2. *O. corniculata*
	Stipules absent	3. *O. radicosa*
4.	Flowers solitary; lateral leaflets usually oblique	4. *O. obliquifolia*
	Flowers in pseudumbels or dichotomous cymes	5
5.	Vertical rhizome present, the bulbs at some depth below the soil-surface	6
	Vertical rhizome absent, the bulbs at or just below the soil-surface, with the leaves and peduncles arising directly from them	11
6.	Leaflets entire or shallowly emarginate	7
	Leaflets deeply lobed or forked at apex	10
7.	Leaflets obovate in outline, with the lateral nerves incurving and confluent near the apex	5. *O. anthelmintica*
	Leaflets linear-lanceolate, lanceolate, ovate to elliptic, or if obovate then lateral nerves not confluent	8
8.	Calyx and pedicels without glandular hairs	6. *O. oligotricha*
	Calyx and pedicels glandular hairy	9
9.	Leaflets membranous, lanceolate-oblong	7. *O. goetzei*
	Leaflets thick, obovate to elliptic	8. *O. trichophylla*
10.	Leaflets forked at apex, oblong to linear-lanceolate	9. *O. abercornensis*
	Leaflets deeply lobed, obcordate in outline	10. *O. semiloba*
11.	Leaflets obtriangular, the lobes usually acute and spreading; flowers in pseudumbels	11. *O. latifolia*
	Leaflets obcordate, the lobes rounded and sometimes overlapping; flowers usually in dichotomous cymes	12. *O. corymbosa*

1. **O. barrelieri** *L.*, Sp. Pl., ed. 2: 624 (1762); Jacq., Oxal.: 24, t. 3 (1794); Knuth in E.P. IV. 130: 65, fig. 5/A (1930). Type: presumably fig. 1139 in Barreliero, Pl. Gall. Hisp. Ital. (1714); no specimen in Herb. Linn.

Erect branched herb, often woody at the base, 15–60 cm. high, densely leafy from the base. Leaves alternate or fasciculate, pinnately 3-foliolate; leaflets lanceolate to ovate-lanceolate, narrowed at both ends, glabrous above and hairy beneath; terminal leaflet the largest, 13–55 mm. long and 9–31 mm. wide, with a stalk up to 10 mm. long, the lateral ones 12–35 mm. long and 5–24 mm. wide, subsessile; petiole about as long as the terminal leaflet, tomentose and not widening into a stipule at the base. Inflorescence cymose, determinate at first, becoming bifid and indeterminate with branches up to

20 mm. long, but sometimes contracted and umbelliform; peduncles axillary near ends of stem and branches, 4–7 cm. long; bracts linear, 1–1·5 mm. long, acute, pubescent; pedicels 2–5 mm. long, jointed at 1 mm. from the base. Sepals linear-lanceolate, 3·5–5 mm. long, acute, membranous, puberulous. Petals white, pink or purplish, with 2 yellow spots at the base, spathulate, cuneate, 7–8 mm. long. Stamen-tube 1·5 mm. long; long filaments dentate in the middle, puberulous; short filaments not dentate, glabrous. Ovary oblong, glabrous; styles longer than both stamen whorls, or intermediate between them in length and hirsute, or shorter and glabrous. Capsule ovoid to oblong, ending bluntly at each end, up to 11 mm. long and 4·5 mm. in diameter. Seeds 3 per locule, 2·5 mm. long, flattened, ± 4-angled, acuminate, rugose.

UGANDA. Mengo District: Kampala, Oct. 1937, *Chandler* 1989! & Sept. 1930, *Hansford* in *Snowden* 1787!

TANGANYIKA. Lushoto District: Amani, 2 Jan. 1950, *Verdcourt* 16! & Sigi, 21 Sept. 1960, *Paulo* 818!; Tanga District: Mlingano, Nov. 1956, *Sandford* 25!

DISTR. **U4**; **T3**; Ivory Coast, Cameroun, probably introduced from tropical America and the West Indies

HAB. In forest clearings, also a weed of roadsides and cultivated ground; 900–1220 m.

SYN. *Lotoxalis barrelieri* (L.) Small in North Amer. Fl. 25: 49 (1907)

2. **O. corniculata** *L.*, Sp. Pl.: 435 (1753); Sond. in Fl. Cap. 1: 351 (1860); P.O.A. C: 225 (1895), pro parte; Knuth in E.P. IV. 130: 146 (1930), pro parte; Salter in Journ. S. Afr. Bot., Suppl. Vol. 1: 73 (1944); C.F.A. 1: 261 (1955); A.V.P.: 126, 293 (1957); Wilczek in F.C.B. 7: 7 (1958), pro parte; F.P.U.: 64, fig. 19 (1962), pro parte; Exell in F.Z. 2: 150 (1963), pro parte. Type: Herb. Linn. No. 600.31 (LINN, lecto.—see Eiten in Taxon 4: 101 (1955))

Annual or perennial much branched herb 10–30 cm. high. Stems creeping, sometimes ascending, frequently rooting at the nodes. Leaves alternate, digitately 3-foliolate, with adnate stipules at the base; stipules 1·5–2 mm. long, ciliate or rarely glabrous; leaflets sessile, ± equal, obcordate, with a narrow sinus about one-third of way down, cuneate, pilose, particularly beneath and at margins. Flowers in 1–6-flowered pseudumbels; peduncles axillary, 1–8 cm. long, about as long as the petioles, pubescent or glabrous, ascending; bracts 2–many, linear-lanceolate, 2–4 mm. long, acute to acuminate, pilose; pedicels 7–15 mm. long. Sepals ovate-lanceolate, 2–5 mm. long, hairy. Petals yellow, shallowly emarginate, cuneate, 4–10 mm. long. Filaments glabrous, the long ones as long as the sepals. Styles hairy, about as long as the long filaments. Capsule linear-oblong, usually abruptly narrowing at apex, (4–)9–17 mm. long, puberulous. Seeds 1–many per locule, ovoid-ellipsoid, flattened, brown, transversely ridged. Fig. 1/5, p. 4.

var. **corniculata**

Stems creeping and ascending. Leaflets up to over 15 mm. long and 20 mm. wide, asually thinly punctulate only beneath. Capsules mostly much over 5 mm. long.

UGANDA. Kigezi District: Kachwekano farm, May 1949, *Purseglove* 2795!; Mengo District: Old Entebbe, 21 Jan. 1956, *Harker* 144!

KENYA. NE. Elgon, Aug. 1963, *Tweedie* 2674!; Nairobi, 26 Aug. 1949, *Jeffery* 655!; Kericho District: SW. Mau Forest, Sambret Catchment, 18 July 1962, *Kerfoot* 3919!

TANGANYIKA. Moshi District: Lyamungu, 10 Sept. 1941, *Wallace* 951!; Uluguru Mts., Morningside, 2 Aug. 1951, *Carmichael* 75!; Songea District: 5 km. E. of Songea by R. Luhira, 15 Jan. 1956, *Milne-Redhead & Taylor* 8332!

ZANZIBAR. Zanzibar, Residency Garden, 12 Apr. 1961, *Faulkner* 2806!; Pemba I., Chake Chake, 24 Aug. 1929, *Vaughan* 573!

DISTR. **U**1–4; **K**2–7; **T**1–4, 6–8; **Z**; **P**; widespread in most tropical and many temperate countries

HAB. Weed in cultivation, disturbed ground, lawns and roadsides; 10–2950 m.

FIG. 1. *OXALIS ANTHELMINTICA*—**1,** habit, × $\frac{2}{3}$; **2,** longitudinal section of flower, × 3. *O. SEMILOBA* subsp. *SEMILOBA*—**3,** leaflets, showing variation in shape, × 1. *O. SEMILOBA* subsp. *UHEHENSIS*—**4,** same, × 1. *O. CORNICULATA*—**5,** insertion of petiole, showing stipule, × 5. *O. RADICOSA*—**6,** insertion of petiole, × 5. 1, from *Hedberg* 1958; 2, from *Mooney* 8917; 5, from *Drummond & Hemsley* 2893; 6, from *E. Brown* 29.

VARIATION. This is a very variable species, in mode of growth, size and pubescence of its parts. There are forms with purplish and variegated leaves.

var. **repens** (*Thunb.*) *Zucc.* in Denkschr. Akad. Wiss. Muench., Abh. 1: 230 (1831). Types: several specimens in Herb. Thunb., one apparently from the Cape Province of South Africa collected by Thunberg (UPS, syn.)

Stems much branched, prostrate, rooting almost at every node. Leaflets very small, up to 4 mm. long and 5 mm. wide, with very sparse long hairs, thickly punctulate on both surfaces. Capsules almost globose, very short, 3–4 mm. long, rarely longer.

UGANDA. Kigezi plains and lower slopes of Mt. Muhavura, 17 Oct. 1929, *Snowden* 1525!; Elgon, Jan. 1918, *Dummer* 3517!
KENYA. Naivasha District: Kinangop, Lereko [Loreko], 21 Dec. 1930, *Napier* 713! & N. Aberdare Mts., Feb. 1936, *Meinertzhagen* in *Herb. Amani* 9342!; Mt. Kenya, *Le Pelley* in *Bally* 3498!
TANGANYIKA. Masai District: Ngorongoro Crater, Apr. 1941, *Bally* 2338!; Kilimanjaro, Marangu, Oct. 1893, *Volkens* 1138!; Mbeya, 1 July 1951, *Bullock* 3987!
DISTR. **U**2, 3; **K**3, 4, 6; **T**2, 7; widespread
HAB. Upland forest glades and moor; 2400–4100 m.

SYN. *O. repens* Thunb., Diss. Oxal.: 16, t. 11 (1781)
O. procumbens A. Rich., Tent. Fl. Abyss. 1: 123 (1847). Types: Ethiopia, Enschecap, *Schimper* II, 1165 (P, syn., BM, K, isosyn.!) & Ouodgerate, *Petit* & Simen [Semien], Mt. Buahit, *Schimper* (both P, syn.)
O. corniculata var. *procumbens* (A. Rich.) Engl. in E.J. 19: 33 (1894) & P.O.A. C: 225 (1895)

NOTE. Whether this variant of *O. corniculata* which occurs on the high mountains of tropical Africa is rightly associated with var. *repens* or whether it should be segregated as var. *procumbens*, as suggested by Engler (see synonymy above), requires a detailed study of the species beyond the scope of this Flora. It seems preferable to adopt the older and more familiar name at present.

3. **O. radicosa** *A. Rich.*, Tent. Fl. Abyss. 1: 123 (1847). Type: Ethiopia, Ouodgerate, *Petit* 80 (P, holo.!, K, iso.!)

Perennial herb, with usually thin stems arising from a thick tuberous main rootstock 5 mm. or more across. Stems erect or ascending, not rooting at the nodes, somewhat angular, loosely pubescent with spreading hairs, more densely so at the nodes and on young parts. Leaves alternate or rarely fasciculate, very like those of *O. corniculata* but with no stipules at the base of the petiole; leaflets very variable in size, usually softly hairy on both sides, sparsely so on the upper side, ciliate at margins. Peduncles axillary, usually 1–1·5(–2·5) times as long as the subtending petioles; bracts 2–many, variable, linear to oblanceolate, ciliate; pedicels 1–2 cm. long, with short spreading hairs which sometimes appear curly on dry specimens. Sepals linear-lanceolate, up to 5 mm. long, often tinged with purple, softly hairy and ciliate at apex. Petals usually appearing pinkish in bud, sometimes with reddish markings in the throat and greenish-yellow at the base, 10 mm. long or rarely longer, glabrous, sometimes ciliate at the top. Short filaments glabrous or rarely hairy; long filaments glabrous or pilose, as long as the sepals or longer. Styles 5, as long as the long filaments or intermediate in length between the 2 filament whorls or shorter than both whorls (specimens from Moroto), hirsute; stigmas laciniate. Capsule cylindrical, gradually tapering to the persistent styles, 5-lobed, up to 3 cm. long, closely pubescent with hairs usually pointing backwards. Seeds many in each locule, brown, somewhat flattened and transversely ridged. Fig. 1/6.

UGANDA. Karamoja District: Mt. Moroto, Feb. 1936, *Eggeling* 2893!; Teso District: Serere, Dec. 1931, *Chandler* 291!; Mengo District: Entebbe, Mar. 1904, *E. Brown* 29!
KENYA. NE. Elgon, Aug. 1963, *Tweedie* 2675!; Uasin Gishu District: near Kapsaret [Kaposoret] Forest Reserve, 8 May 1951, *G. R. Williams* 169!; Nairobi, near National Museum, 11 June 1952, *Kirrika* 181!
TANGANYIKA. Ngara, 12 Dec. 1959, *Tanner* 4623!; W. slopes of Kilimanjaro, Ol Molog, 21 June 1944, *Greenway* 6910!; Iringa District: Iheme, 28 Jan. 1934, *Michelmore* 935!

DISTR. **U1–4**; **K1, 3–7**; **T1–7**; N. Nigeria east to the Somali Republic and south to Mozambique, Malawi and Zambia, also India

HAB. Forest clearings, grassland, rocky slopes, a weed on cultivated ground and roadsides; 300–2100(–2400) m.

SYN. [*O. corniculata* L. var. *stricta* sensu Oliv. in F.T.A. 1: 297 (1868), quoad descr. et specim. cit., *non* (L.) Oliv. sensu stricto]
[*O. stricta* sensu R. E. Fries, Schwed. Rhod.-Kongo-Exped. 1: 107 (1914); F.P.N.A. 1: 401 (1948), *non* L.]
[*O. corniculata* sensu Knuth in E.P. IV. 130: 146 (1930), pro parte; W.F.K.: 22 (1948), pro parte; Wilczek in F.C.B. 7: 7 (1958), pro parte; F.P.U.: 64 (1962), pro parte; Exell in F.Z. 2: 150 (1963), pro parte, *non* L.]

NOTE. As may be seen from the synonymy, *O. radicosa* has usually been confused with *O. corniculata* L., and those authors such as Robyns, in F.P.N.A., who saw that two species were present confused it with *O. stricta* L., whose type (seen in the British Museum), like *O. corniculata*, possesses stipules. Where *O. radicosa* has been confused with *O. corniculata* it is usually impossible from the descriptions alone to decide which species is referred to.

The species is very variable as may be seen from the description. It shows, unlike *O. corniculata*, a high degree of morphological (and ? probably functional) heterostyly such as is found in many other members of the family. Specimens from Karamoja are usually compact and have both sets of filaments very hairy.

4. **O. obliquifolia** *A. Rich.*, Tent. Fl. Abyss. 1: 123 (1847); Knuth in E.P. IV. 130: 348 (1930); Salter in Journ. S. Afr. Bot., Suppl. Vol. 1: 154 (1944); C.F.A. 1: 262 (1951); Brenan in Mem. N.Y. Bot. Gard. 8: 231 (1953); Wilczek in F.C.B. 7: 8 (1958); Exell in F.Z. 2: 155 (1963). Types: Ethiopia, Tigre, " Dscheladscheranne ", *Schimper* III, 1643 (P, syn., BM, K, isosyn. !) & near Adua, *Quartin Dillon* & Ouodgerate, *Petit* (both P, syn.)

Acaulescent bulbous herb with a vertical rhizome 1–10 cm. long. Bulb solitary at base of rhizome, ovoid-ellipsoid, up to 12 mm. long and 6–12 mm. in diameter, covered by 2 or more dirty brown reticulate scales. Leaves 3-foliolate, appearing in a rosette at the soil-surface, with petioles widening at the base; leaflets sessile, broadly obovate, cuneate, the middle one usually shallowly emarginate, slightly wider than the lateral ones, 5–11 mm. long and 9–19 mm. wide, and with a distinct midvein, the lateral ones 5–11 mm. long and 5·5–16 mm. wide, with no midvein but with up to 4 veins arising from the base, usually foveolate on both sides, glabrous or rarely ciliate on the midvein below. Flowers solitary on erect peduncles longer than petioles; bracts 2, linear, 1·5–2(–4) mm. long. Sepals oblong-lanceolate, sometimes with 2 apical calli, 3–4(–5) mm. long, shortly hairy and ciliate at the top, usually tinged purple. Petals pink or purple, sometimes with a yellow base, narrowly obovate, 10–23 mm. long, entire, hairy in part on the outside. Filaments glabrous. Styles densely hairy; flowers can be long-styled (styles longer than both stamen whorls), mid-styled (styles intermediate in length between the 2 stamen whorls) or short-styled (styles shorter than both stamen whorls), but the homostylous condition (styles as long as long filaments) has not been noticed; stigmas laciniate. Capsule globose, shorter than the sepals, 3–3·5 mm. long, terminated by the persistent styles. Seeds 1 or 2 in each locule, 1 mm. long, light brown with faint markings.

UGANDA. Acholi District: Imatong Mts., Apr. 1938, *Eggeling* 3538 !; Karamoja District: Kokumongole, 28 May 1939, *A. S. Thomas* 2861 !

KENYA. Turkana District: Napau Hills, 10 May 1953, *Padwa* 110 !; Nairobi District: Mbagathi, Bahati Estate, Dec. 1932, *C. G. Rogers* 201 !

TANGANYIKA. Mbulu District: Mbulumbulu, 23 June 1944, *Greenway* 6919 !; Kahama District: Kahama–Biharamulo road 35 km. E. of Ushirombo, 27 Jan. 1962, *Boaler* 458 !; Mbeya Mt., 13 May 1956, *Milne-Redhead & Taylor* 10212 !

DISTR. **U1**; **K1–6**; **T1–8**, Ethiopia, Sudan, Congo, Zambia, Rhodesia, Malawi, Angola, Mozambique and South Africa

HAB. Open or wooded grassland and *Brachystegia* woodland, often in moist places on sloping hillsides or amongst rocks, also a weed of cultivated ground and roadsides; 600–3170 m.

SYN. *O. obliquifolia* A. Rich. var. *eminii* Engl., P.O.A. C: 225 (1895). Type: Tanganyika, Kilosa District, Kimamba [Kimambiro], *Stuhlmann* 110 (B, holo. †, BM, K, iso. !)
O. obliquifolia A. Rich. var. *transvaalensis* Knuth in E.P. IV. 130: 348 (1930); Suesseng. in Trans. Rhod. Sc. Assoc. 43: 1071 (1951). Types: South Africa, Transvaal, Pretoria, *Wilms* 197 & Lydenburg, *Wilms* 195 (both B, syn. †)

NOTE. Reference to this species as a " little devil " by Jex-Blake, W.F.K.: 22 (1948), most probably refers to 11, *O. latifolia*.

5. **O. anthelmintica** *A. Rich.*, Tent. Fl. Abyss. 1: 124 (1847); Salter in Journ. S. Afr. Bot., Suppl. Vol. 1: 86 (1944); Wilczek in F.C.B. 7: 9 (1958); Exell in F.Z. 2: 151 (1963). Type: Ethiopia, Tigre, valley of R. Tacazze near " Dscheladscheranne ", *Quartin Dillon* (P, holo.)

Acaulescent or rarely a shortly stemmed perennial herb, with a vertical rhizome up to 18 cm. long arising from the top of an ovoid 1–1·5 cm. long bulb. Leaves 3-foliolate, usually appearing in a rosette but sometimes spaced and alternate; petioles widening at the base, either glabrous or with long spreading hairs; leaflets sessile, subcircular to obovate, rounded or shallowly emarginate at apex, cuneate, glabrous above and with a few hairs at least on the midrib beneath, sometimes gland-dotted at the edge. Flowers pink, purplish or blue in pseudumbels; peduncles arising from the leaf-rosette or subtended by cauline bracts, up to over 20 cm. long and usually longer than the petioles; bracts variable, short and linear to elongate and lanceolate, rarely with apical calli; pedicels up to 2 cm. long, with spreading hairs. Sepals ovate-lanceolate, 4–7 mm. long, acute, sparsely hairy, with or without orange apical calli. Petals up to 16 mm. long, hairy in part on the outside or glabrous. Stamens with long filaments pubescent and short filaments glabrous. Styles pubescent, intermediate in length between or longer than the 2 stamen whorls. Capsule globose, shorter than the sepals, with 3 seeds in each locule. Seeds shiny, reddish-brown, rugose. Fig. 1/1, 2, p. 4.

UGANDA. Acholi District: Imatong Range, Langia Mts., Apr. 1943, *Purseglove* 1387 !; Karamoja District: Mt. Moroto, May 1954, *M. S. Philip* 558 !
KENYA. Northern Frontier Province: Moyale, 16 Apr. 1952, *Gillett* 12821 !; eastern slope of Elgon, near Suam R., 1 May 1948, *Hedberg* 810 !; Nairobi District: Mbagathi, 16 May 1948, *Bally* 6230 !
TANGANYIKA. Moshi, Mar. 1927, *Haarer* 455 ! & Moshi–Arusha, near Kikafu bridge, 15 Mar. 1955, *Huxley* 161 ! & Kilimanjaro, Kibosho [Kiboscho], Mar. 1894, *Volkens* 2055 !
DISTR. **U**1; **K**1–4, 6; **T**2; Ethiopia, Congo, Zambia, Rhodesia, Malawi and Mozambique
HAB. Occasional weed in grassland and at edge of riverine forest, on sandy loam, appearing after rains; 840–2130 m.

SYN. *O. abyssinica* Turcz. in Bull. Soc. Nat. Mosc. 31: 432 (1858), *non* (A. Rich.) Walp. (1852), *nom. illegit.* Type: Ethiopia, Tigre, near Adua, Mt. Scholoda, *Schimper* 188 (CWU, holo., K, iso. !)
O. caprina L. var. *abyssinica* (Turcz.) Oliv. in F.T.A. 1: 296 (1868)
O. caprina L. var. *anthelmintica* (A. Rich.) Engl., P.O.A. C: 225 (1895)
O. purpurata Jacq. var. *anthelmintica* (A. Rich.) Knuth in E.P. IV. 130: 304 (1930)

NOTE. *O. anthelmintica* A. Rich. var. *glanduligera* Engl. is most likely to be *O. trichophylla* Bak., see p. 9.

6. **O. oligotricha** *Bak.* in K.B. 1895: 64 (1895); Knuth in E.P. IV. 130: 302 (1930); Exell in F.Z. 2: 152, t. 23/C (1963). Type: Zambia, near Mbala [Abercorn], Fwambo, *Carson* (K, holo. !)

Perennial herb, acaulescent or sometimes with a distinct aerial stem up to 15 cm. long, with a vertical rhizome up to 12 cm. long arising from the top of the bulb; bulb up to 1·3–4 cm. long and 1·2 cm. in diameter, the scales broadly ovate-lanceolate, acute to acuminate, brown, smooth with parallel veins; rhizome with membranous scales 2–3 mm. long, which continue as green cauline bracts and sometimes subtend the peduncles. Leaves 3-foliolate, in a rosette and either basal or high up on the stem, long-petiolate; leaflets very variable in form and size, broadly ovate, obovate or ovate-lanceolate to lanceolate, acute to obtuse at apex, almost glabrous or sparsely hairy above and softly so beneath, ciliate at the margin. Flowers purple or blue, in pseudumbels; peduncle 10–15 cm. long; bracts linear and variable in length, up to 6 mm. long, pilose; pedicels 1–2 cm. long, with spreading hairs or glabrous. Sepals oblong-lanceolate, 4–5 mm. long, acute with 2 apical calli, pilose. Petals oblanceolate, 9–13 mm. long, glabrous. Filaments glabrous. Styles pubescent, as long as, or longer than, the long-stamen whorl, thus the flowers are either long-styled or homostyled. Capsule globular, with 3 seeds in each locule. Seeds flattened-ellipsoid, longitudinally ribbed, brown.

TANGANYIKA. Njombe District: Msima Stock Farm, *Emson* 338!; Songea District: E. of Songea by Nonganonga stream, 27 Dec. 1955, *Milne-Redhead & Taylor* 7916! & about 16 km. W. of Songea, 2 Jan. 1956, *Milne-Redhead & Taylor* 8108!
DISTR. T7, 8; Mozambique (*fide* Exell), Zambia
HAB. Upland grassland and in *Brachystegia* woodland; 960–1900 m.

7. **O. goetzei** *Engl.* in E.J. 28: 413 (1900) & V.E. 3 (1): 716, fig. 329 (1915); Knuth in E.P. IV. 130: 301, fig. 19 (1930). Type: Tanganyika, Iringa District, Muhinde Steppe, *Goetze* 511 (B, holo. †)

Acaulescent perennial herb, with a simple vertical rhizome, the rhizome often with membranous scales getting larger towards the top; bulb not seen. Leaves 3-foliolate, 3–many in a rosette; petioles up to over 13 cm. long, with a mixed indumentum of sparse long simple hairs and short glandular hairs; leaflets linear-oblong, up to more than 35 mm. long and 12 mm. wide, narrowing to the base and obtuse at apex, glabrous above, punctate and sparsely hairy beneath, more hairy on the midrib. Flowers 5–many in pseudumbels, light purple with a greenish throat; peduncles as long as, or more often longer than, the petioles, arising from the leaf-rosette or subtended by cauline bracts, sparsely hairy or glabrous; bracts many, shortly lanceolate to linear, 4–5 mm. long, acute, pilose; pedicels 1–2 cm. long, like the calyx with glandular hairs. Sepals lanceolate, 4–7 mm. long, acute with apical calli, glandular hairy. Petals cuneate and spathulate, ± 3 times as long as the sepals, hairy in part on the outside or glabrous. Long filaments hairy, ± 1·5 times as long as the sepals; short filaments glabrous. Styles shorter or longer than the 2 stamen whorls, thus the flowers are short-styled or long-styled. Capsule globular, with up to 3 seeds in each locule. Seeds reddish-brown.

TANGANYIKA. Dodoma District: Saranda Escarpment, 15 Dec. 1931, *B. D. Burtt* 3504!; Mpwapwa District: Kongwa, 22 Jan. 1950, *Anderson* 577! & Mpwapwa, 11 Dec. 1934, *Mr. & Mrs. Hornby* 629!
DISTR. T5, 7; not known elsewhere
HAB. *Brachystegia-Julbernardia* woodland on slopes of escarpments; 1150–1400 m.

8. **O. trichophylla** *Bak.* in K.B. 1895: 63 (1895); Knuth in E.P. IV. 130: 302 (1930); Exell in F.Z. 2: 151 (1963). Type: Zambia, near Mbala [Abercorn], Fwambo, *Carson* 56 (K, holo.!)

Acaulescent or very shortly stemmed perennial herb, with a vertical rhizome up to 7 cm. long; bulb ovoid, with a linear-lanceolate beak, up to 3 cm. long excluding the beak and 2 cm. in diameter; beak 1–2·5 cm. long; scales brown, smooth, with parallel veins. Leaves 3-foliolate, 2–many ± in a basal rosette; petiole up to 11 cm. long, pubescent; leaflets obovate to elliptic, very variable in form and size, from 1·5–4(–5) cm. long and up to

2 cm. wide above the middle, obtuse, rounded or sometimes truncate at apex, cuneate, sparsely pilose to glabrate above, densely pilose beneath, ciliate at the margins. Flowers in 2–many-flowered pseudumbels, pink to purplish-blue with a yellowish throat; peduncles longer than the petioles, up to 15 cm. long, arising from the leaf-rosettes or subtended by cauline bracts below the rosettes, with simple and glandular hairs; bracts linear, up to 6 mm. long; pedicels up to 1·5 cm. long, glandular hairy. Sepals broadly lanceolate, 6·5 mm. long, glandular hairy, with or without orange apical calli. Petals oblong, obliquely obtuse at apex, cuneate, 10–15 mm. long. Long filaments hairy, with short appendages at the back in the lower half, in mid- and long-styled flowers short filaments glabrous and styles hairy, in short-styled flowers short filaments hairy and styles glabrous. Ovary globose. Capsule glabrous or hairy, with 2–3 seeds per locule. Seeds up to 1·5 mm. long and 1·0 mm. wide, longitudinally 6–7-ribbed, chocolate-brown.

TANGANYIKA. Ufipa District: E. of Kito Mt., 12 Dec. 1934, *Michelmore* 1052! & base of Kito Mt., Chito village, 2 Jan. 1968, *Richards* 22856!
DISTR. **T4**, 7; Zambia
HAB. Cultivated ground on sandy soil; 1800 m.

NOTE. *O. anthelmintica* A. Rich. var. *glanduligera* Engl., in E.J. 28: 412 (1900), based on *Goetze* 529 (B†) from Kigonsive near Iringa and cited by Knuth in E.P. IV. 130: 304 (1930) under *O. purpurata* var. *anthelmintica*, is more probably *O. trichophylla* which sometimes has leaves resembling those of *O. anthelmintica* but has glandular hairs on the pedicels and calyx.

9. **O. abercornensis** *Knuth* in F.R. 48: 3 (1940); Exell in F.Z. 2: 152 (1963). Type: Tanganyika, Ufipa District, SE. of Kito Mt., *Michelmore* 1049 (B, holo. †, EA, K, iso. !)

Perennial herb, acaulescent or nearly so, with a vertical rhizome up to 20 cm. long, but usually very short, arising from the top of an elongated bulb; bulb 2–5 cm. long, 0·6–1·3 cm. in diameter, with smooth parallel-veined brown scales. Leaves 3-foliolate, 2–5, ± basal; petioles up to 20 cm. long, widening at the base for 1 cm. or more, pubescent to densely pilose; leaflets sessile, linear to oblong, widening to the deeply forked apex, 4–7 cm. long, 0·3–0·7 cm. wide, glabrescent above and hairy beneath, with clear venation, not obviously foveolate. Flowers purple, in 4–25-flowered pseudumbels, borne on peduncles up to 25 cm. long, which arise from the leaf-rosette or from the axils of bracts below, pilose. Bracts many, linear, up to 6 mm. long; pedicels up to 23 mm. long. Sepals lanceolate, 4·5–8 mm. long, acute, sparsely hairy, with orange apical calli. Petals up to 12(–14) mm. long, glabrous. Long filaments hairy; short filaments glabrous. Styles either longer than both stamen whorls (long-styled) and hairy or shorter than both (short-styled) and then glabrous. Capsule shortly oblong, up to 5 mm. long, shorter than the sepals or sometimes as long. Seeds up to 3 in each locule, reddish-brown, flattened and faintly transversely ridged.

TANGANYIKA. Ufipa District: SE. of Kito Mt., 12 Dec. 1934, *Michelmore* 1049!
DISTR. **T4**; known only from the type-locality and one place in Zambia between Kito Mt. and Mbala [Abercorn]
HAB. Along paths and as a garden weed on sandy soil; 1650–1800 m.

NOTE. From information given by Michelmore, the above is the precise type-locality, which was not known by Knuth or Exell.

10. **O. semiloba** *Sond.* in Fl. Cap. 1: 350 (1860); Oliv. in F.T.A. 1: 296 (1868); Salter in Journ. S. Afr. Bot., Suppl. Vol. 1: 87 (1944); Wilczek in F.C.B. 7: 10 (1958); Exell in F.Z. 2: 153, t. 23/A (1963). Types: South Africa, Orange Free State, Caledon R., *Burke* 440 & *Zeyher* 271 & Cape Province, Somerset, *Atherstone* (all K, isosyn. !)

Perennial herb, with a vertical rhizome arising from an ovoid 1–4 cm. long bulb; aerial stem usually present, up to 15 cm. long, sometimes very short or rarely absent, hairy, with 1–3 alternate ovate-lanceolate hairy cauline bracts up to 10 mm. long. Leaves 3-foliolate, 2 to many, mostly in a rosette at apex of stem (rarely alternate) or basal when the stem is very short; petioles 5–16 cm. long; leaflets sessile, obovate in outline, bilobed with lobes varying from rounded to oblong, cuneate, nearly glabrous to sparsely hairy, especially beneath, usually foveolate. Flowers pinkish to purple, in pseudumbels; peduncles up to 17 cm. long, arising from the leaf-rosette or subtended by cauline bracts below; bracts few to many, linear, 2–5(–7) mm. long, acute, sometimes with apical calli. Pedicels 1–2(–2·8) cm. long, sparsely hairy, becoming glabrous. Sepals ovate, obtuse, 3·5–5 cm. long, with apical calli. Petals spathulate, cuneate, 10–15 mm. long. Long filaments hairy, usually as long as the sepals; short filaments about one-third the length of the long ones. Styles hairy; flowers heterostylous, long-styled, mid-styled or short-styled. Capsule globose, shortly hairy, terminated by the persistent styles. Seeds 2 or rarely 3 in each locule.

subsp. **semiloba**

Leaflets shallowly to deeply obcordate or broadly emarginate to lobed half-way down; lobes convex on their apical margins so that they make a sharp angle between them. Fig. 1/3, p. 4.

TANGANYIKA. Njombe District: Lupembe, Mar. 1931, *Schlieben* 428!; Lindi District: Nachingwea, 20 Jan. 1952, *Anderson* 742!
DISTR. **T**2, 7, 8; Mozambique west to Angola and southwards to South Africa
HAB. Upland grassland and *Brachystegia* woodland in fairly moist sites; 450–2740 m.

subsp. **uhehensis** (*Engl.*) *Exell* in Bol. Soc. Brot., sér. 2, 35: 135 (1961) & in F.Z. 2: 153 (1963). Type: Tanganyika, Iringa District, edge of Uhehe Plateau near Lula [Sula], *Goetze* 499 (B, holo. †)

Leaflets broadly obcordate to deeply 2-lobed or 2-partite; lobes narrowly oblong to oblong-lanceolate, concave on their apical margins so that they make a wide ± rounded angle between them. Fig. 1/4, p. 4.

TANGANYIKA. Kondoa District: Kikore [Kikori], 10 Mar. 1930, *B. D. Burtt* 2710!; Chunya District: Lupa N. Forest Reserve, 29 Dec. 1962, *Boaler* 802!; Songea District: Matengo Hills, E. of Ndenga [Ndengo], 13 Jan. 1956, *Milne-Redhead & Taylor* 8242!
DISTR. **T**5, 7, 8; Congo, Zambia, Rhodesia, Malawi
HAB. *Brachystegia* woodland and upland grassland, on red or sandy soils; 840–1670 m.

SYN. *O. uhehensis* Engl. in E.J. 28: 412 (1900) & V.E. 3(1): 716, fig. 330 (1915); Knuth in E.P. IV. 130: 309, fig. 20 (1930)
O. angustiloba R. E. Fries, Schwed. Rhod.-Kongo-Exped. 1: 107 (1914); Wilczek in F.C.B. 7: 10 (1958). Types: Zambia, near Mbala [Abercorn], Msisi, *Fries* 1298 & Kalambo, *Fries* 1389 (both UPS, syn.)

11. **O. latifolia** *Kunth*, Nov. Gen. et Sp. Pl. 5: 237, t. 467 (1822); Knuth in E.P. IV. 130: 273 (1930); Salter in Journ. S. Afr. Bot., Suppl. Vol. 1: 73 (1944); Wilczek in F.C.B. 7: 11 (1958); Exell in F.Z. 2: 155 (1963). Type: Mexico, near Campeche, *Humboldt* (P, holo.)

Acaulescent herb, with leaves and peduncles arising directly from the bulb; bulb ovoid, 7–15 mm. in diameter, with 1 or more thick roots up to 2 mm. wide and many other fibrous ones at the base; scales papery, but some of the outer ones striated and thickened with a starchy substance; stolons with membranous scales arise from the bulb and develop at the end a bulbil which eventually forms a new plant. Leaves 3-foliolate; leaflets widely obtriangular, up to 5 cm. long and 8 cm. wide, cuneate, shallowly emarginate at the apex so that the lobes are widely diverging, green, paler beneath, sparsely hairy or glabrate when mature with only a few hairs along the margins and midrib. Flowers light purple with a green throat, in 5- to over 20-flowered pseudumbels borne on long peduncles, which usually exceed the petioles; bracts 2, widely ovate, 2 mm. long and 1·5–2 mm. wide, acute, papery, ciliate; pedicels up to 2 cm. long, glabrous. Sepals oblong-lanceolate, 3·5–5 mm. long, acute, each with 2 orange apical calli which sometimes turn darkish brown in dried specimens. Petals falsely connate at the base, spathulate, up to 3 times as long as the sepals. Long filaments 5–6 mm. long, hairy in the upper half with 1 mm. long appendages on the outside; short filaments sparsely hairy or more often, glabrous. Styles shortly hairy, shorter than the 2 stamen whorls, thus flowers are short-styled; stigmas laciniate. Capsule elongate and glabrous; mature capsules not seen.

UGANDA. Kigezi District: Kachwekano Farm, June 1950, *Purseglove* 3472!; Mengo District: near Kampala, 3 Apr. 1961, *Miller* 464!

KENYA. Uasin Gishu District: Lamok R. on Ol Dane Sapuk road, 3 May 1951, *G. R. Williams* 151!; Kilifi District: Kibarani, 9 May 1946, *Jeffery* 527!

TANGANYIKA. Moshi District: Lyamungu, 25 May 1945, *Wallace* 1250! & Mar. 1957, *Ivens* 877!; Morogoro District: Kibuko Old Mica Camp, Mar. 1955, *Mgaza* 12!

DISTR. **U**2–4; **K**3–5, 7; **T**1–3, 6; Congo, Sudan, Ethiopia, Zambia, Rhodesia, Mozambique, South Africa; native of America, introduced in Africa

HAB. Cultivated ground and waste places, usually in moist and well-shaded parts; sea-level to 2520 m.

NOTE. A very common and obnoxious weed, very difficult to eradicate because of its formation of many bulbils.

Although the oldest herbarium specimen from East Africa seems to be *Webster* in *Herb. Amani* 8736, collected in 1942 at Kitale, there is a painting in the East African Herbarium prepared in Kenya in January 1933 and the species is referred to in Jex-Blake et al., Gardening in East Africa, ed. 1 (1934), as thriving under warm conditions and to be avoided as a potential weed. It has often been confused with indigenous species, especially *O. semiloba* and, to a lesser extent, *O. anthelmintica* and *O. obliquifolia*, and the reputation of these species as bad weeds is largely or wholly due to this fact. Fig. 89 in Ivens, E.A. Weeds and their Control (1967) wrongly referred to *O. latifolia* in fact represents *O. semiloba* subsp. *semiloba*.

12. **O. corymbosa** *DC.*, Prodr. 1: 696 (1824); Exell, Cat. Vasc. Pl. S. Tomé: 124 (1944); Wilczek in F.C.B. 7: 9 (1958); Exell in F.Z. 2: 155 (1963). Types: Mauritius, *Bory* & Réunion, collector unrecorded (G-DC, syn.)

Bulbous acaulescent perennial herb; bulb rounded, 15–25 mm. in diameter, made up of many bulbils up to 5 mm. in diameter borne on a central axis; bulb-base corm-like with many fibrous roots; scales oblong-lanceolate, brown, papery, thicker at the 3 mid-nerves, the inner scales thickened with a starchy substance. Leaves 3-foliolate, on long petioles up to 25 cm. long; leaflets sessile, obcordate, bilobed, up to 4 cm. long and 4·5 cm. wide; lobes rounded, hairy beneath, ciliate, sometimes with orange dots near the margins. Flowers pink, purplish or blue, greenish in the throat in a pseudumbel or a bifid branching cyme on long hairy or glabrate peduncles usually longer than the petioles; bracts 2 at base of inflorescence and then 1 at the base of each branch, ovate, subacute, ciliate with 1 or 2 linear calli on the midrib; pedicels hairy, up to 3 cm. long. Sepals oblong-lanceolate, 4·5–5 mm. long, subacute, hairy and each with 2 orange calli near the apex. Petals obovate, cuneate, ± 3 times as long as the sepals, rounded at the apex. Long filaments hairy, 6–7 mm. long; short filaments almost glabrous, 2–3 mm. long. Styles very hairy, about as long as the sepals, between the 2 stamen whorls in length, thus the flowers are mid-styled; stigmas laciniate, blackish in dry specimens. Capsule elongate; mature capsule not seen.

TANGANYIKA. Moshi District: Lyamungu, May 1958, *Ivens* 1140!; E. Usambara Mts., Amani, 14 June 1939, *Greenway* 5882!

DISTR. **T**2, 3; native of tropical America, a long-standing weed of cultivation in the Old World tropics

HAB. In shaded parts of cultivated ground and waste places; 800–1980 m.

SYN. *O. martiana* Zucc. in Denkschr. Akad. Muench. 9: 144 (1825); Knuth in E.P. IV. 130: 250 (1930). Types: Brazil, near Rio de Janeiro, *Martius & Beyrich* (M, syn.)

NOTE. The first specimen of this species in East Africa seems to have been collected in 1913 by Frau Braun at Amani (in *Herb. Amani* 5233). It has until now been confused with *O. anthelmintica*.

2. BIOPHYTUM

DC., Prodr. 1: 689 (1824)

Annual or perennial herbs, with a simple or branched stem, or stem very short and then plants with a thick woody rootstock. Leaves fascicled or almost whorled at the top of the stem or branches, paripinnate (the rhachis ending in a bristle), exhibiting a certain degree of sensitivity to touch; petiole swollen and articulated near the base; leaflets many, opposite or subopposite, subsessile or shortly petiolulate, obliquely rectangular or orbicular, unequal-sided. Flowers in pseudumbels, white, pink, yellow or orange; bracts small, persistent, forming an involucre at the base of the pedicels. Sepals strongly parallel-veined, the inner ones in relation to the pseudumbel narrower than the outer. Petals contorted in bud, becoming shortly adherent above the free base, falling off together in a bunch. Stamens 10, 5 outer short and 5 inner long; filaments shortly connate or almost free at the base. Ovary 5-lobed; styles notched or 2-fid at apex. Capsule ovoid or oblong, dehiscing loculicidally, the valves finally spreading. Seeds tuberculate, rugose or ribbed.

A tropical genus of about 70 species.

1. Annual, or perennial with a long slender straight stem 2
   Perennial, acaulescent or almost so, with a thick woody rootstock 5
2. Leaflets membranous, the terminal pair about twice as large as the next pair; sepals 2–2·5 mm. long 1. *B. abyssinicum*
   Leaflets semi-rigid, the terminal pair 1–1·5 times as large as the next pair; sepals 4–6 mm. long 3
3. Lateral nerves thick and prominent, making a right angle with the midrib; sepals 6 mm. long . 4. *B. petersianum*
   Lateral nerves thin, making an acute angle with the midrib; sepals 4–5 mm. long 4
4. Bracts long-acuminate; leaflet-halves at the base ± equal . . . . 2. *B. helenae*
   Bracts subacute; leaflet-halves at the base very unequal, the anterior half much larger . . 3. *B. turianiense*
5. Pseudumbels sessile . . . . 8. *B. macrorrhizum*
   Pseudumbels pedunculate 6
6. Leaflets obliquely rectangular to spathulate-oblong 5. *B. kassneri*
   Leaflets elliptic to orbicular 7
7. Sepals, pedicels and peduncles glandular . . 6. *B. nyikense*
   Sepals, pedicels and peduncles eglandular . . 7. *B. crassipes*

1. **B. abyssinicum** *A. Rich.*, Tent. Fl. Abyss. 1: 122 (1847); Knuth in E.P. IV. 130: 395, fig. 23/A–C (1930); C.F.A. 1: 263, t. 9/A (1951); Delhaye in Inst. Roy. Col. Belg. 23: 855, t. 2/52–61 (1952); F.W.T.A., ed. 2, 1: 59 (1954); Exell in F.Z. 2: 156, t. 24/A (1963). Types: Ethiopia, Tigre, Shire [Chiré], *Quartin Dillon* (P, syn.) & Guendepta [Gapta], *Schimper* 1206 (P, syn., K, isosyn.!)

Annual herb; stem slender, simple, straight, 5–30 cm. high and 1–1·5 mm. in diameter, terete, glabrate or sparsely appressed hairy with downward

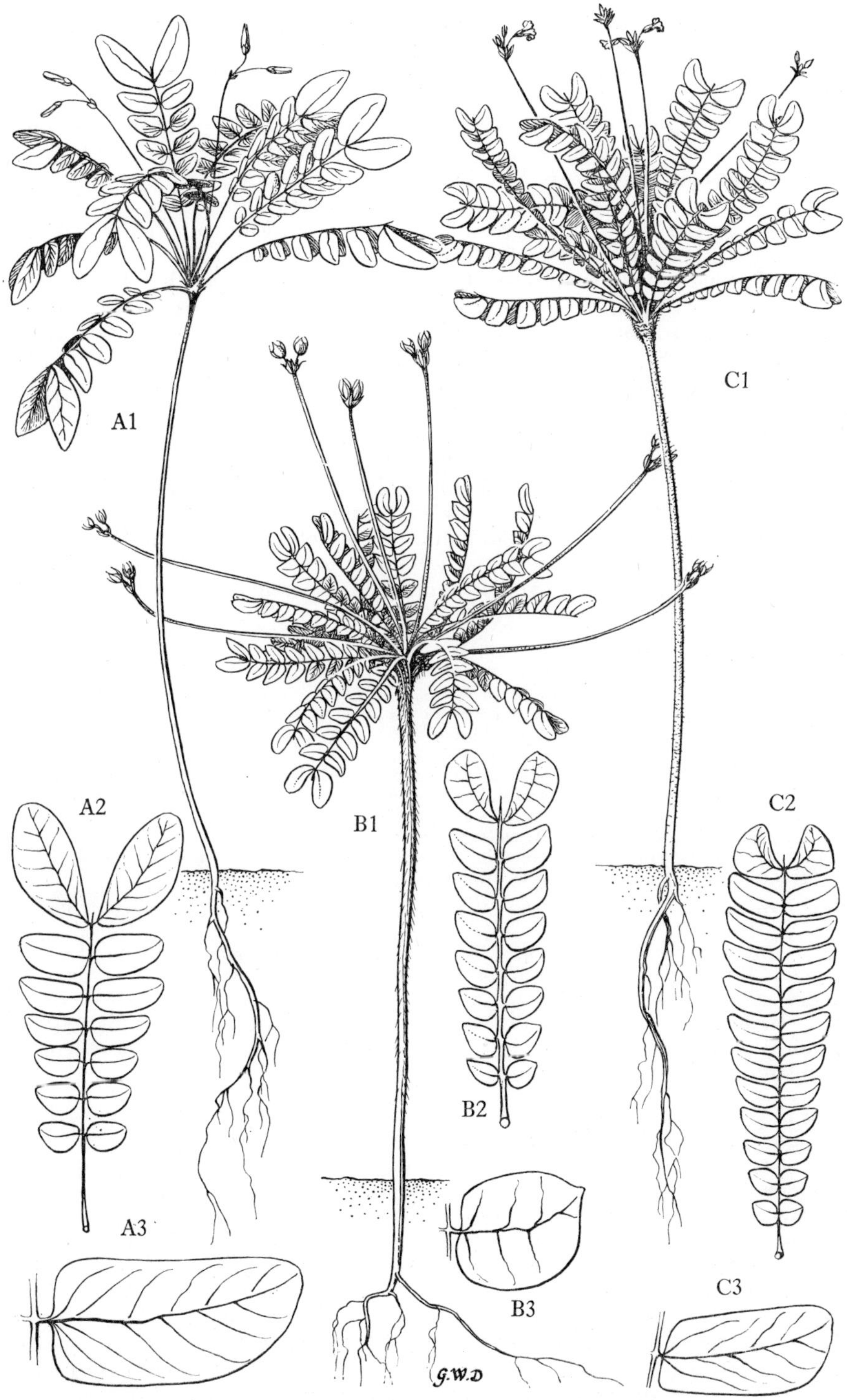

Fig. 2. *BIOPHYTUM ABYSSINICUM*—**A1**, habit, × ½; **A2**, leaf, × 1; **A3**, leaflet, × 3. *B. PETERSIANUM*—**B1**, habit, × ½; **B2**, leaf, × 1; **B3**, leaflet, × 3. *B. HELENAE*—**C1**, habit, × ½; **C2**, leaf, × 1; **C3**, leaflet, × 3. A1–3, from *Chase* 4388; B1–3, from *Chase* 814; C1–3, from *Exell, Mendonça & Wild* 1357. Reproduced by permission of the Editors of Flora Zambesiaca.

directed hairs. Leaves 5 to many in a simple rosette at the top of the stem, up to 7 cm. long and 12–16 mm. wide, 3–11(usually 7)-jugate, sensitive to touch, the leaflets closing together upwards; petioles and rhachis often thin, glabrous or slightly hairy; leaflets subsessile or with a short petiolule 0·5 mm. long, green or tinged with purple, the apical leaflets the largest, almost twice the size of the next pair, obliquely elliptic, up to 20 mm. long and 9 mm. wide, obtuse at apex, narrowly cuneate, the median leaflets oblique- to oblong-rectangular, the basal ones very small, up to 2 mm. long and 2 mm. wide; all membranous, paler green beneath and with up to 10 pairs of very clear nerves, glabrous or sometimes ciliate. Flowers in peduncled 1–5(–12)-flowered pseudumbels; peduncles slender, as long as or longer than the leaves, 2–7(–10·5) cm. long, glabrous or sparsely appressed hairy; bracts very small, linear, 1–1·5 mm. long, acute, 1-nerved; pedicels up to 7 mm. long, glabrous or pubescent. Sepals lanceolate, acutely acuminate, 2–2·5 mm. long, 3-nerved, often membranous at the margins. Petals free, later adherent for two-thirds of their length above the free bases, spathulate, 2–2·5 times as long as the sepals, 5–6(–17) mm. long, white, creamy yellow or pinkish-mauve with a yellow base. Filaments nearly free to base, the long filaments hairy, the short ones glabrous. Styles of varying length in relation to the filaments, the flowers thus trimorphic heterostyled; stigmas bifid. Capsule subglobose to obovoid, up to 4 mm. long and 3 mm. in diameter, longer than the sepals, darker and hairy at ridges, with 1–3 seeds per locule. Seeds flattened-ellipsoid, up to 1·3 mm. long and 0·7 mm. wide, tuberculate, brown. Fig. 2/A, p. 13.

UGANDA. W. Nile District: Mt. Otzi [Otze], Oct. 1959, *E. M. Scott* in *E.A.H.* 11795!; Bunyoro District: Budongo Forest, Busingiro rest house, 13 Aug. 1962, *Miller* 590!; Mengo District: Kipayo, Jan. 1914, *Dummer* 577!

KENYA. Northern Frontier Province: Moyale, 3 Sept. 1953, *Bally* 9105!; Masai District: Emali Hill, 27 Dec. 1960, *Archer* 231! & Chyulu foothills, 19 Apr. 1938, *Bally* 83!

TANGANYIKA. Shinyanga District: Mwantini [Mantini] Hills, 30 Mar. 1932, *B. D. Burtt* 3767!; Moshi, June 1927, *Haarer* 326!; Dodoma District: 35 km. S. of Itigi Station on Chunya road, 17 Apr. 1964, *Greenway & Polhill* 11598!

DISTR. **U**1–4; **K**1, 3, 4, 6; **T**1–8; widespread in tropical Africa

HAB. In shady often rocky places in open or wooded grassland and in *Brachystegia* woodland on red sandy soil; 380–1600 m.

SYN. *Oxalis abyssinica* (A. Rich.) Walp. in Ann. Bot. Syst. 2: 241 (1852); Oliv. in F.T.A. 1: 297 (1868)

Biophytum reinwardtii (Zucc.) Klotzsch subsp. *abyssinicum* (A. Rich.) Steen. in Bull. Gard. Bot. Buitenz., ser. 3, 18: 453 (1950); Wilczek in F.C.B. 7: 13 (1958)

2. **B. helenae** *Buscal. & Muschl.* in E.J. 49: 475 (1913); Veldkamp in Blumea 16: 137 (1968). Type: Zambia, between Lakes Bangweulu and Tanganyika, *Duchess of Aosta* 1154 (B holo. †)

Annual herb; stem generally simple but sometimes shortly branched at apex, usually slender but sometimes up to 5 mm. in diameter, ± widening at the apex, hairy. Leaves in 1 or, in stout plants, up to 5 rosettes at apex of stem or branches, 2–15 cm. long, 5–21-jugate, sensitive to touch, the leaflets closing together upwards; leaflets subsessile, the apical ones usually the largest, 1–1·5 times the length of next pair, decreasing in size downwards, the largest obliquely obovate, the others obliquely rectangular, rounded at the apex but sometimes abruptly and shortly apiculate, truncate at the base, pubescent on both surfaces or glabrous except at the ciliate margin; lateral nerves ± inconspicuous from above, thin but distinct from beneath, usually twice as many in the lower half as in the upper half (halves in relation to leaf-rhachis); petiole and rhachis pubescent. Flowers in peduncled 3–10-flowered pseudumbels; peduncles usually longer than the leaves, pubescent; bracts 2 mm. long, acutely acuminate, 1-nerved, pubescent; pedicels up to 7 mm.

long, pubescent, sometimes glandular hairy, articulate up to 3 mm. above the point of insertion, so that when the upper parts drop off the basal stumps are seen among the bracts. Sepals 4–5 mm. long, acutely acuminate, 3–5-nerved, pubescent, sometimes also with a few glandular hairs. Petals spathulate, 1·5–2 times as long as the sepals, white, yellow or pinkish-purple, glabrous. Filaments shortly connate at the base. Ovary deeply 5-lobed; styles hairy near the base, glabrous towards the apex and shortly bifid. Capsule obovoid, 3 mm. long and 2 mm. in diameter. Seeds flattened-ellipsoid, small, up to 0·5 mm. long, rugose, brown. Fig. 2/C, p. 13.

Uganda. Ruwenzori, Bwera, Oct. 1940, *Eggeling* 4102!; Kigezi District: Kayonza Forest Reserve, 2 Aug. 1960, *Paulo* 630!
Tanganyika. Buha District: Kalinzi, 4 July 1963, *Carmichael* 916!; Rungwe District: Tukuyu, Nov. 1928, *Haarer* 1667!; Njombe District: Mkinji R., Mar. 1954, *Semsei* 1681!
Distr. **U**2, 4; **T**4, 7, 8; Rwanda, Burundi, Congo, Zambia, Malawi, Angola, Mozambique
Hab. Riversides and in clearings of rain-forest and riverine forest, less commonly in open or wooded grassland; 790–1980 m.

Syn. [*Oxalis sensitiva* sensu Oliv. in F.T.A. 1: 297 (1868), pro parte, *non* L.]
[*Biophytum sensitivum* sensu Hook., Niger Fl.: 269 (1849); Klotzsch in Peters, Reise Mossamb. Bot. 1: 85 (1862); Knuth in E.P. IV. 130: 393 (1930); Delhaye in Inst. Roy. Col. Belg. 23: 850, t. 1/1–9 (1952); Wilczek in F.C.B. 7: 14 (1958); Exell in F.Z. 2: 158, t. 24/C (1963), *non* (L.) DC.]

Note. True *B. sensitivum* (L.) DC. is restricted to the Indo-Pacific region as shown recently by Dr. J. F. Veldkamp in the reference cited.

3. **B. turianiense** *Kabuye* in K.B. 23: 297, fig. 1/1 (1969). Type: Tanganyika, Morogoro District, Turiani, *Semsei* 1432 (EA, holo.!, K, iso.!)

Annual or perennial; stem single, sometimes up to 5-branched at the top, straight, 15–25 cm. long, sometimes ascending and rooting in the lower parts, becoming thickened and normally rough with many leaf-scars on the upper part on older plants, shortly pilose, the hairs becoming denser towards the top. Leaves in a single rosette at top of stem or branches, up to 10 cm. long, 3–10 (–18)-jugate; petiole up to 15(–18) mm. long, long-hirsute towards the base; rhachis pubescent; leaflets obliquely oblong, rectangular, acute to abruptly and shortly apiculate, truncate at the base except for the terminal ones which are almost cuneate, the terminal ones the largest, 12–14(–23) mm. long, 4–6(–11) mm. wide, pubescent (sometimes sparingly so) beneath, glabrous above except sometimes along the midrib and at the ciliate margin; lateral nerves thin but distinct beneath, numerous, in up to 13 or more pairs, almost parallel, making an acute angle with the midrib. Peduncles ± as long as the leaves, closely hairy; bracts many, making a globular or oblong head 4–5 mm. in diameter, ovate-lanceolate, 1·5–2 mm. long, subacute, 1-nerved; pedicels up to 6 mm. long, pilose. Sepals lanceolate, 4–4·5 mm. long, acute, 3–6-nerved, sparsely and shortly hairy. Corolla 5·5 mm. long, cream, glabrous; petals at first free, afterwards falsely shortly connate above the free pointed bases, rounded at the apex. Filaments connate for 0·5 mm. at the base; short filaments 1–2 mm. long and glabrous, almost white; long filaments 2·5–3·5 mm. long and hairy, darker coloured. Ovary globose, deeply 5-lobed; styles longer than the 2 filament whorls or shorter and then deflexed, joined for 0·5 mm. to form a stylar tube, hairy with club-shaped hairs in the upper half, widening into a flattened stigmatic surface. Capsule globular or ± obovoid, overtopped by remains of styles. Seeds 3 in each locule, 1·0–1·3 mm. long and 0·7 mm. wide, reddish-brown, longitudinally ribbed.

Tanganyika. Morogoro District: Turiani, Nov. 1953, *Semsei* 1432! & 3 June 1933, *B. D. Burtt* 4729! & Uluguru Mts., Mkungwe Hill, 5 July 1970, *Faden* in *Kabuye* 286!
Distr. **T**6; not known elsewhere

HAB. Riverine and dry evergreen forest; 600–900 m.

NOTE. This species is near *B. helenae* but can easily be separated from it by characters given in the key.

4. **B. petersianum** *Klotzsch* in Peters, Reise Mossamb. Bot. 1: 81, t. 15 (1861); C.F.A. 1: 264, t. 9/D (1951); Brenan in Mem. N.Y. Bot. Gard. 8: 231 (1953); F.W.T.A., ed. 2, 1: 159, fig. 59 (1954); Wilczek in F.C.B. 7: 15, t. 1 (1958); Exell in F.Z. 2: 158, t. 24/B (1963). Types: Zanzibar, *Peters* & Mozambique, Querimba, *Peters* (both B, syn. †)

Annual; stem simple, straight, (4–)10–20(–40) cm. high, 2 mm. in diameter, with downward pointing hairs becoming denser towards the top. Leaves subsessile, in a single rosette 2–7 cm. in diameter, up to 2–7 cm. long and 6–12 mm. wide, 3–8(–10)-jugate; rhachis hirsute; terminal leaflets the largest, up to 6–10 mm. long and 3–6 mm. wide, obliquely obovate, ± 1·5 times the size of the next pair, the middle leaflets rectangular-ovate, the basal ones ovate-triangular, rounded at the base, sparsely long hairy or glabrous except at the ciliate margin, paler beneath and with prominent nerves, the lateral ones making a right-angle with the midrib and bifurcating towards the margins. Flowers white, yellow, orange or reddish, in generally sessile but sometimes pedunculate 2–6-flowered pseudumbels; peduncles, when present, up to 5·5 cm. long and longer than the leaves, ascending, appressed hairy; bracts linear-lanceolate, 2·5 mm. long, acute to long-acuminate, 1-nerved, usually hairy; pedicels slender, 3–4 mm. long, articulate at about 2 mm. from the point of insertion, hirsute especially towards the top. Sepals linear-lanceolate, 4–6 mm. long, 1 mm. wide, acute, 5–7-nerved, hirsute to glabrate. Petals 1·5 times as long as the sepals, 5–8 mm. long, retuse, narrowly cuneate, connate in the middle to form a tube above the free bases, glabrous. Filaments almost free at the base; long filaments hairy, the short ones glabrous. Ovary deeply 5-lobed, globose and hairy; styles of varying length in relation to the stamens, making the flowers trimorphic heterostylous, hairy; stigma laciniate. Capsule 3 mm. long and 2 mm. in diameter, pubescent or glabrate. Seeds up to 5 per locule, small, rounded, ± 0·5 mm. in diameter, largely transversely ridged, light brown. Fig. 2/B, p. 13.

UGANDA. W. Nile District: Mt. Otzi [Otze], Oct. 1959, *E. M. Scott* in *E.A.H.* 11796!; Kigezi District: Ruhinda, Jan. 1951, *Purseglove* 3531!; Mengo District: Old Entebbe, 28 Jan. 1956, *Harker* 154!

KENYA. N. Kavirondo District: about 13 km. E. of Bungoma, 27 June 1955, *Bogdan* 4042!; Kericho District: Cheptuiyet, Belgut, 28 Aug. 1960, *Kerfoot* 2144!; Kilifi District: Sokoke Forest, 5 Oct. 1945, *Jeffery* 342!

TANGANYIKA. Mwanza District: between Geita and Karumo, 31 Mar. 1937, *B. D. Burtt* 6462!; W. Usambara Mts., Kwamshemshi–Sakare road, 4 July 1953, *Drummond & Hemsley* 3167!; Bagamoyo District: Kidomole, Apr. 1964, *Semsei* 3750!

ZANZIBAR. Zanzibar I., Zingwe Zingwe, 21 Jan. 1929, *Greenway* 1107!; Pemba I., Madunga, 29 July 1929, *Vaughan* 494!

DISTR. **U**1–4; **K**5, 7; **T**1, 3, 4, 6–8; **Z**; **P**; widespread in tropical Africa, Madagascar and Asia

HAB. *Brachystegia* woodland, open grassland and marsh edges, or more usually a weed of cultivated ground on sandy soils; sea-level to 1800 m.

SYN. *Oxalis apodiscias* Turcz. in Bull. Soc. Nat. Mosc. 36: 595 (1863). Types: Java, *Horsfield* (CWU, syn.) & India, Assam, Khasia Mts., *Hooker & Thomson* 5 (CWU, syn., K, isosyn.)

[*O. sensitiva* sensu Oliv. in F.T.A. 1: 297 (1868), pro parte, *non* L.]

Biophytum apodiscias (Turcz.) Edgew. & Hook.f. in Fl. Br. India 1: 437 (1874); W.F.K.: 22 (1948)

Oxalis sessilis Baill. in Bull. Soc. Linn. Paris 1: 598 (1886), *nom. illegit.* Type as for species

Biophytum sessile (Baill.) Knuth in E.P. IV. 130: 406 (1930); Delhaye in Bull. Inst. Roy. Col. Belg. 23: 852, t. 1/11–14, 16–20, 24, 25, 31 (1952), *nom. illegit.*

B. rotundifolium Delhaye in Bull. Inst. Roy. Col. Belg. 23: 855, t. 1/22, 23, 28, 29 (1952). Type: Congo, Katanga, Kapiri valley, *Homblé* 1107 (BR, holo.!)

5. **B. kassneri** *Knuth* in E.P. IV. 130: 409 (1930); Wilczek in F.C.B. 7: 18 (1958); Exell in F.Z. 2: 161, t. 25/D (1963). Type: Congo, Katanga, Mt. Senga, *Kassner* 2983a (B, holo. †, BR, K, iso. !)

Acaulescent perennial herb, with a thickened woody hairy rootstock; rootstock not more than 2 cm. long and 12 mm. in diameter, truncate, shortly divided, with many leaves crowded at the top; taproot simple, ± 3 mm. in diameter. Leaves almost sessile, 6–17 cm. long and 12–15 mm. wide, 8–16-jugate; rhachis densely hairy to subhirsute; leaflets subsessile, ± rectangular-oblong, 2–12 mm. long, 2–6 mm. wide, rounded to subacute at the apex, widely cuneate to rounded at the base, pubescent on both surfaces and margin or glabrate, prominently nerved beneath with the ascending lateral nerves making an acute angle with the midrib. Flowers white, yellow or pink to mauve with orange centre, in pedunculate, 3–10-flowered pseudumbels; peduncles as long as or longer than the leaves, subhirsute; bracts ovate-lanceolate, 1–2 mm. long, acuminate, 1-nerved, pubescent; pedicels slender, 1·5–4(–10) mm. long, articulated above the point of insertion, pubescent. Sepals oblong-lanceolate, 3–4·5 mm. long, acute to shortly acuminate, 3–5-nerved, sparsely pubescent. Petals about twice as long as the sepals, glabrous. Filaments shortly connate at the base; short filaments up to 1·5 mm., the long ones 3–4 mm. long. Ovary 5-lobed; styles in flowers seen are intermediate in length between the 2 filament whorls, thus flowers are mid-styled. Capsule ellipsoid, 2·2 mm. long, 1·8 mm. across. Seeds flattened-ellipsoid, 0·8 mm. long, brown, tuberculate. Fig. 3/D, p. 18.

TANGANYIKA. Ufipa District: Sumbawanga, above Chapota, 11 Mar. 1957, *Richards* 8626!
DISTR. **T4**; Congo (Katanga), Zambia
HAB. Rocky ground in *Brachystegia* woodland; 1740 m.

6. **B. nyikense** *Exell* in Bol. Soc. Brot., sér. 2, 35: 137 (1961) & in F.Z. 2: 161, t. 25/B (1963). Type: Zambia, Nyika Plateau, *Benson* 145 (BM, holo. !)

Perennial herb with a thickened woody rootstock. Leaves in 1(–3) rosettes ± at ground-level, 2–12 cm. long, 4–10-jugate, sensitive; rhachis pubescent; leaflets obliquely subcircular, 2–8 mm. long, 2–7·5 mm. wide, the largest in the middle, rounded to bluntly pointed at the apex, truncate at the dase, with many lateral nerves which form an acute angle with the midrib, glabrous except for the ciliate margin. Flowers yellow, in 3–8-flowered pedunculate pseudumbels; peduncles short, 1–3·5 cm. long, glandular pubescent. Sepals narrowly lanceolate, 6–7 mm. long, 6–8 nerved, glandular pubescent. Petals spathulate, 7–8 mm. long, glabrous. Filaments almost free to the base; short filaments 1 mm. long, the long ones 2·5 mm. long. Ovary 5-lobed; styles pilose, longer than or in between the 2 filament whorls, thus flowers are long-styled or mid-styled. Seeds flattened-ellipsoid, up to 1·3 mm. long, brown. Fig. 3/B, p. 18.

TANGANYIKA. Kigoma District: Mahali Mts., Kabesi, 1 Sept. 1958, *Newbould & Jefford* 2101!; Chunya District: Mbeya–Mwanza, Mbogo, 17 Oct. 1932, *Geilinger* 3108!
DISTR. **T4, 7**; Zambia, Malawi
HAB. *Brachystegia-Uapaca* woodland (also occurs in upland grassland south of the Flora area); ± 1500 m.

7. **B. crassipes** *Engl.*, P.O.A. C: 226 (1895); Knuth in E.P. IV. 130: 408, fig. 23/F–G (1930); Delhaye in Bull. Inst. Roy. Col. Belg. 23: 858 (1952); Wilczek in F.C.B. 7: 19 (1958); Exell in F.Z. 2: 159, t. 25/A (1963). Type: Tanganyika, Mwanza District, Bukindo [Bugando], *Stuhlmann* 3542 (B, holo. †)

Perennial herb, with a thickened woody brown rootstock; rootstock 1–2 cm. long, 1–1·5 cm. in diameter, truncate, simple or divided to bear up to 4 rosettes of leaves ± at ground-level; taproot vertical, subwoody, up to 2 mm.

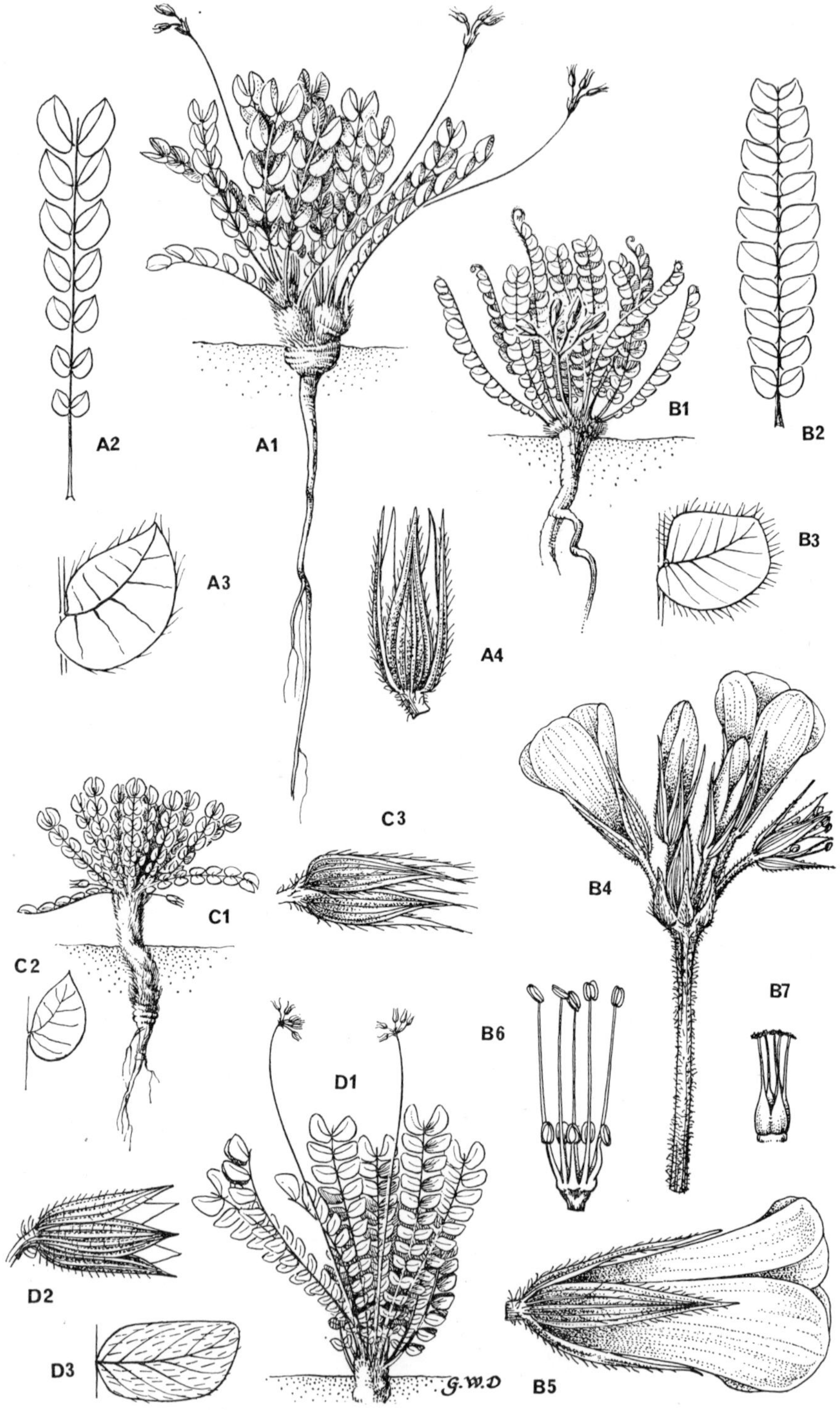

FIG. 3. *BIOPHYTUM CRASSIPES*—**A1,** habit, × ½; **A2,** leaf, × 1; **A3,** leaflet, × 3; **A4,** calyx, × 4. *B. NYIKENSE*—**B1,** habit, × ½; **B2,** leaf, × 1; **B3,** leaflet, × 3; **B4,** inflorescence, × 2; **B5,** flower, × 4; **B6,** stamens, × 5; **B7,** gynoecium, × 5. *B. MACRORRHIZUM*—**C1,** habit, × ½; **C2,** leaflet, × 3; **C3,** calyx, × 4. *B. KASSNERI*—**D1,** habit, × ½; **D2,** calyx, × 4; **D3,** leaflet, × 3. A1–3, from *Kassner* 2022; A4, from *Polhill & Paulo* 1732; B1–4, from *Robson* 207; B5–7, from *Newbould & Jefford* 2101; C1, 2, from *Richards* 1116; C3, from *Richards* 19039; D1–3, from *Fanshawe* 1967. Reproduced in part from Flora Zambesiaca, with the permission of the Editors.

in diameter. Leaves shortly petiolate, 6–18 cm. long and 13–15 mm. wide, 6–17-jugate; rhachis sparsely pilose or glabrate; leaflets obliquely subcircular, obovate or subreniform, 2·5–11 mm. long, 2–7 mm. wide, the terminal ones the largest in mature leaves, rounded at the apex, truncate at the base, glabrous except at the ciliate margin, paler and prominently nerved beneath; lateral nerves in 4–6 pairs, spreading and bifurcating towards the margin. Flowers orange-yellow or reddish, in 4–6-flowered pedunculate pseudumbels; peduncles shorter or slightly longer than the leaves, erect and appressed hairy; bracts linear-lanceolate, 1·5–3 mm. long, long-acuminate, 1–3-nerved, pubescent; pedicels slender, 2–4(–6) mm. long, mostly hidden by the bracts, pubescent. Sepals linear-lanceolate, (3·5–)5–6 mm. long, 5–9-nerved, acute, pubescent. Corolla up to 9 mm. long, glabrous or with white to brown hairs in the throat; petals narrowly cuneate, rounded to retuse at apex. Filaments shortly connate at the base. Ovary deeply 5-lobed; styles hairy, of varying length in relation to the filaments, making the flowers trimorphic heterostylous; stigmas laciniate. Capsule ovoid-ellipsoid, 2–3 mm. long, glabrous. Seeds subglobose, small, ± 0·5 mm. long, transversely tuberculate, brown. Fig. 3/A.

TANGANYIKA. Mwanza District: S. Buhindi Forest Reserve, Sept. 1954, *Carmichael* 442!; Tabora District: Ugalla R., 26 Sept. 1949, *Bally* 7491!; Iringa District: Malakala, 13 Mar. 1962, *Polhill & Paulo* 1732!

DISTR. **T**1, 4, 7; Congo, Zambia, Malawi, Mozambique

HAB. *Brachystegia* woodland and secondary bushland or grassland, sometimes near streams, usually on red sandy soils; 1150–1800 m.

8. **B. macrorrhizum** *R. E. Fries*, Schwed. Rhod.-Kongo-Exped. 1: 108, t. 6/5–6 (1914); Knuth in E.P. IV. 130: 409 (1930); Exell in F.Z. 2: 162, t. 25/C (1963). Type: Zambia, near Mbala [Abercorn], Msisi, *Fries* 1307 (UPS, holo.)

Small perennial herb, with a short woody shortly branched rootstock; branches tufted, 1–3 cm. long and 7–10 mm. wide, densly clothed by leaf-remnants; taproot long, widening upwards. Leaves erect, 2–4 cm. long, 7–10-jugate, the petiole and rhachis pubescent to pilose; leaflets subsessile, asymmetric, obliquely rounded or reniform, the terminal leaflets the largest and up to 5 mm. long, decreasing downwards, the smallest 2 by 2 mm., subacute to acute, sometimes minutely apiculate, truncate at the base, rigid, green, paler beneath and prominently nerved, glabrous except at the ciliate margin; lateral nerves in 4–7 pairs, forming a wide acute angle with the midrib and bifurcating towards the margin. Flowers yellow striated with orange, in subsessile pseudumbels and appearing solitary among the leaf-rosettes; bracts linear-lanceolate, 2–4 mm. long, 1-nerved, pubescent; pedicels slender, 4–10 mm. long, sparsely long-hairy to glabrate. Sepals narrowly ovate-lanceolate, up to 6·5(–8) mm. long, acute, 5–10-nerved, membranous round the margin, laxly pubescent. Petals spathulate, up to 12 mm. long, pointed at the free bases, shortly hairy in the throat inside. Filaments shortly connate at the base; short filaments 3·5 mm. long, glabrous; long filaments 5·5 mm. long, hairy and darker in colour. Ovary oblong-obovoid, 5-lobed, attenuate to the base and long-hairy near apex; styles revolute, glabrous and bifid at the apex, shorter than the 2 filament-whorls, thus flowers are short-styled. Capsule oblong, up to 3 mm. long by 2 mm. across, with 3 seeds per locule. Seeds flattened-ellipsoid, small, up to 0·5 mm. long, closely and deeply tuberculate, reddish-brown. Fig. 3/C.

TANGANYIKA. Kigoma District: 58 km. S. of Uvinza, 31 Aug. 1950, *Bullock* 3285!; Mpanda District: Silkcub Highlands, 8 Dec. 1956, *Richards* 7179!; Ufipa District: Chapota, 4 Dec. 1949, *Bullock* 2011!

DISTR. **T**4; Zambia

HAB. Damp places in *Brachystegia* woodland and associated grasslands; 1220–1730 m.

INDEX TO OXALIDACEAE

GEOGRAPHICAL DIVISIONS OF THE FLORA

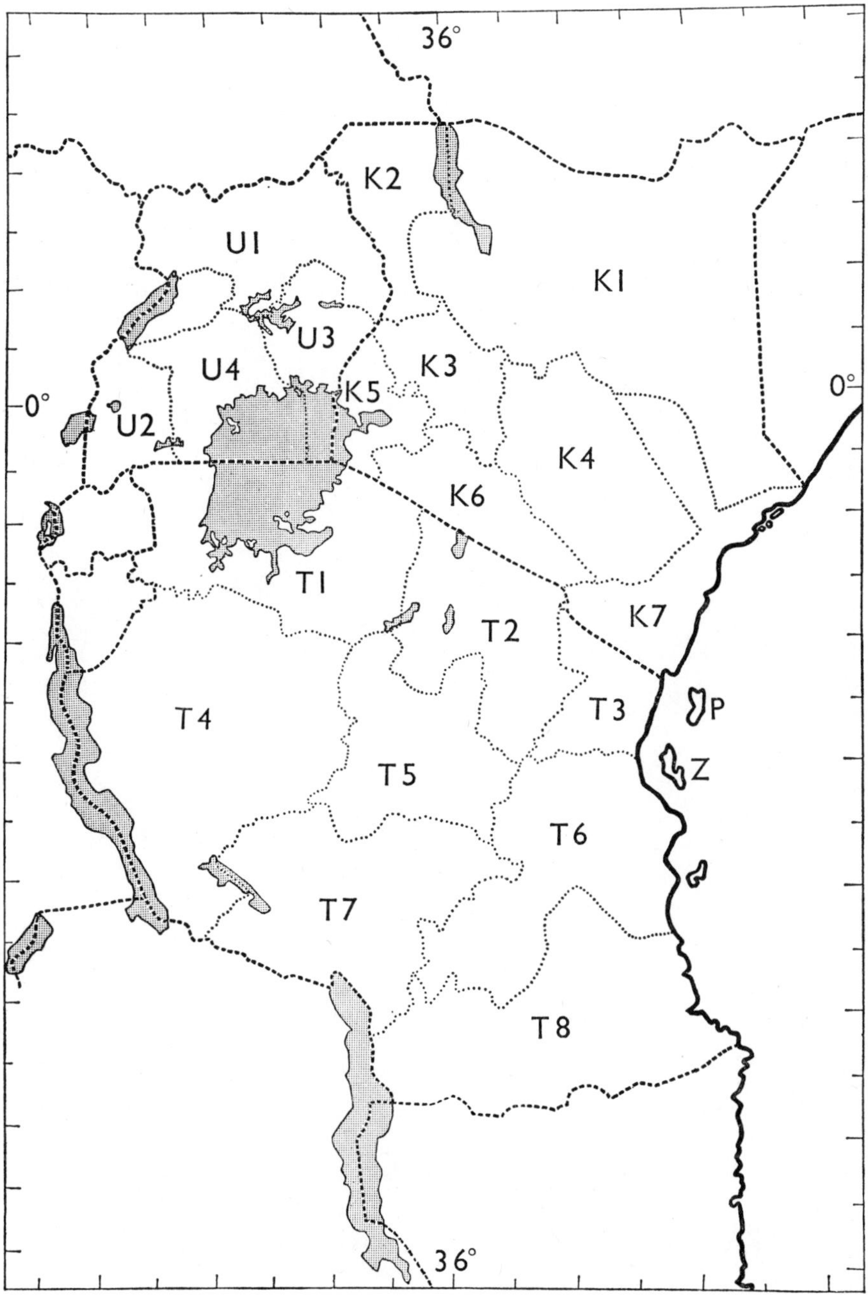